baby Unicorn Activity Book

This book belongs to

....

....

Handwriting Practice

Trace the numbers and letters balow

1 1 1 1 1 1 1 1 1 1

One   One   One

2 2 2 2 2 2 2 2

Two   Two   Two

3 3 3 3 3 3 3 3 3

Three   Three   Three

find the same image and color it

 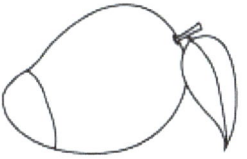

1 One

1 One

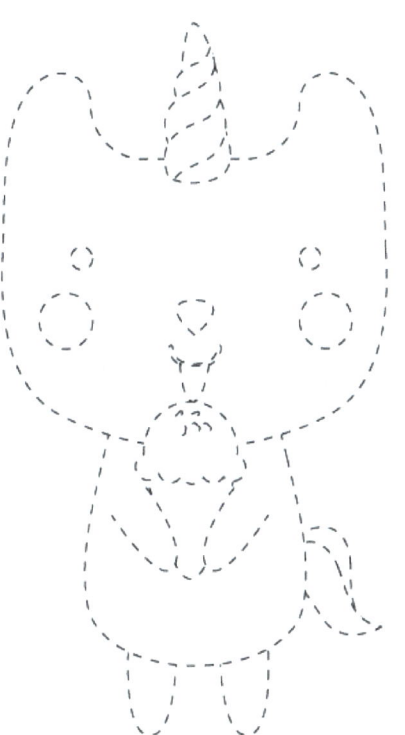

2 Two

2 Two

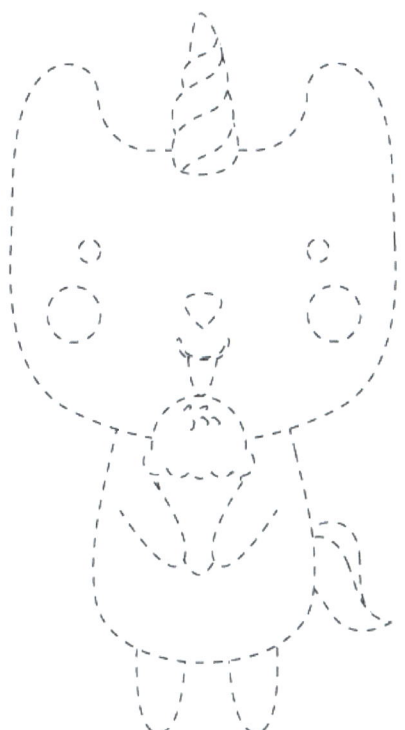
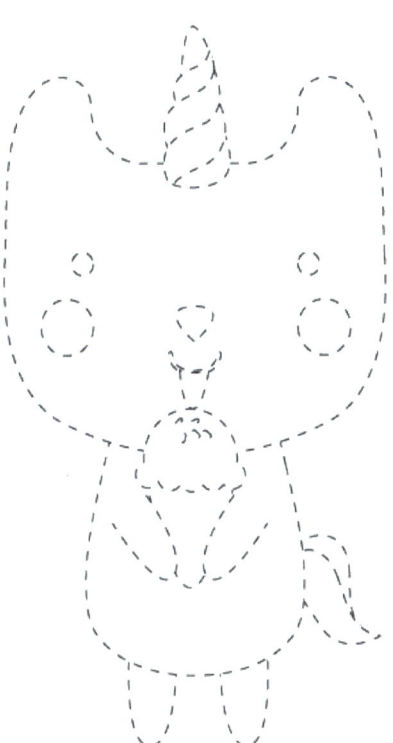

3 Three

3 Three

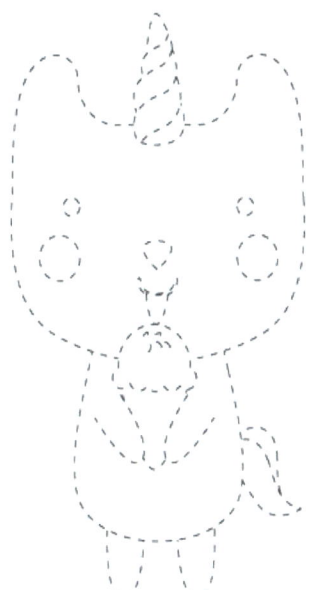
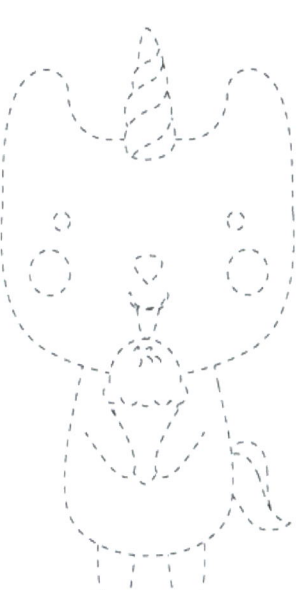

Match the pictures to their shadows

☆

Handwriting Practice

Trace the numbers and letters balow

4 4 4 4 4 4 4 4

Four Four Four

5 5 5 5 5 5 5 5

Five Five Five

6 6 6 6 6 6 6 6

Six Six Six

4 Four

4 Four

5 Five

5 Five

6 Six

6 Six

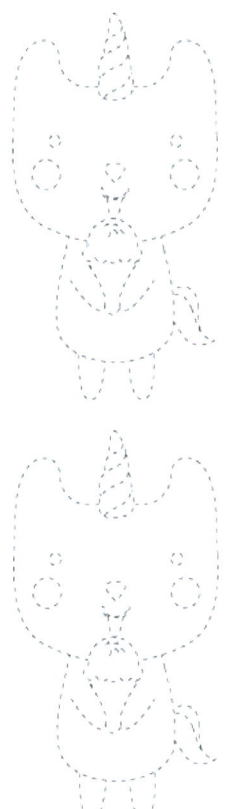
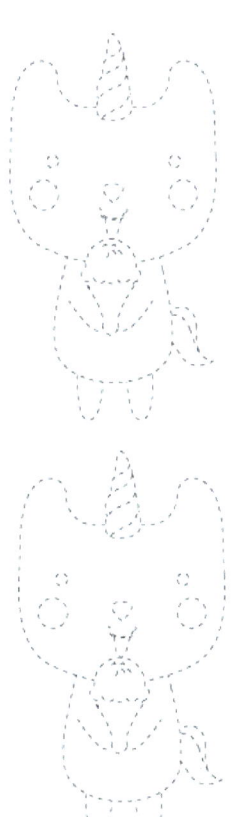
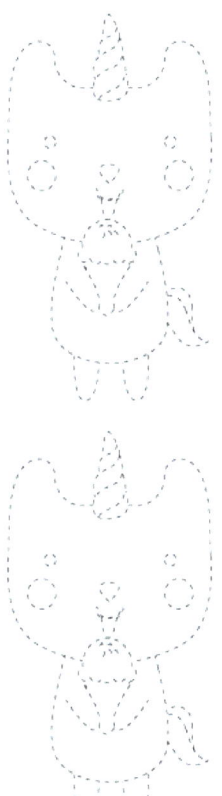

HOW MANY....?

3   4   2

7   8   5

Baby bear dot drawing

Dinosaur coloring

Find the right shade

 A ★
 B ★
 C ★
 D ★

Memory Game

Addition

2 + 1 =

2 + 2 =

Handwriting Practice

Trace the numbers and letters balow

7 7 7 7 7 7 7

Seven Seven Seven

8 8 8 8 8 8 8

Eight Eight Eight

9 9 9 9 9 9 9

Nine Nine Nine

find the same image and color it

7 Seven

7 Seven

Place x in front of the correct

Addition #2

3 + 1 =

4 + 0 =

8 Eight

8 Eight

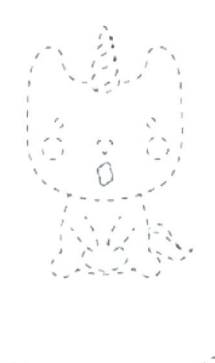

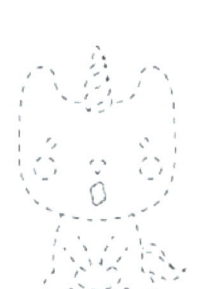
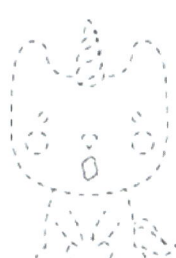

Baby unicorne dot drawing

Find the right shade

 A ★

 ★

 B ★

 ★

 C ★

 ★

 D ★

 ★

9 Nine

9 Nine

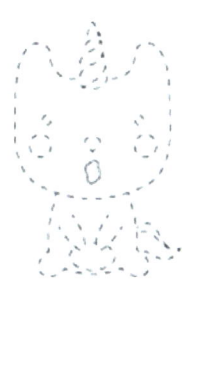 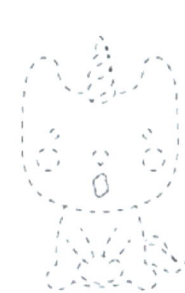 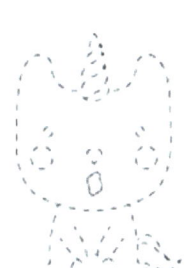

 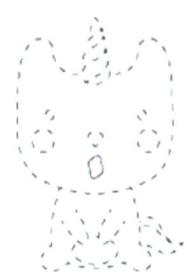 

 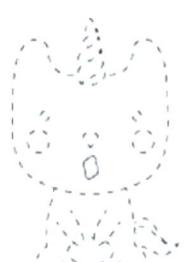 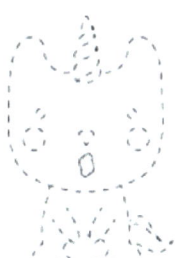

Mouse and cheese coloring

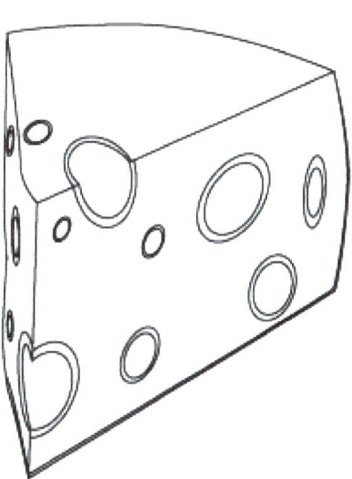

HOW MANY....?

FIND 2 THE SAME PICTURES

Handwriting Practice

Trace the numbers and letters balow

10 10 10 10 10 10

Ten   Ten   Ten

10 Ten

10 Ten

Addition #3

🍅🍅🍅 + 🍅🍅🍅🍅🍅 = 🍅🍅🍅🍅🍅🍅🍅🍅

[3] + [5] = [ ]

- -

🍓🍓🍓🍓🍓🍓 + 🍓🍓 = 🍓🍓🍓🍓🍓🍓🍓🍓

[6] + [2] = [ ]

Cat and Mouse Coloring

 # Keep drawing and coloring the squares

HOW MANY....?

8   7   6

7   8   5

Bear dot drawing

Place x in front of the correct image

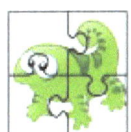

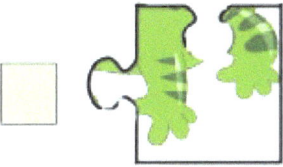

HOW MANY....?

Keep drawing and coloring the trigls

Find the right shade

 A ★   ★

 B ★   ★

 C ★   ★

 D ★   ★

Camel dot drawing

Calculate the addition
and
color the correct number of fruits. #4

3 + 3 =

5 + 4 =

 # Keep drawing and coloring the rectangles

Memory Game

fox and sheep coloring

Lion dot drawing

Place X in front of the correct image

Calculate the addition
and color the correct number of fruits #5

$3 + 4 = \boxed{}$

$4 + 6 = \boxed{}$

 Keep drawing and coloring the stars

Color each shape according to its appropriate color

find the correct shadow

How many shapes...? #1

find the same image and color it

Unicorne dot drawing

How many shapes ? #2

Gazelle dot drawing

Unicorne dot drawing

Solutions

KEY

 A    C

 B    D

 C    A

 D    B

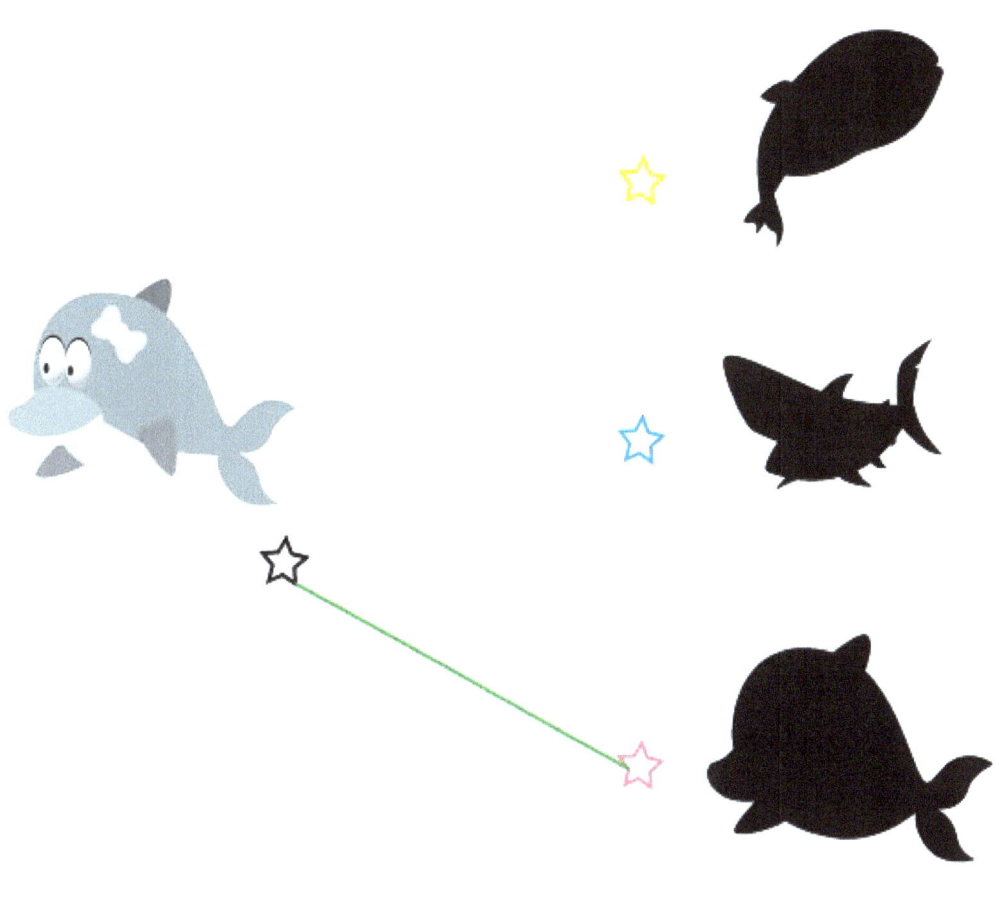

HOW MANY....?

 A ★

 C ★

 B ★

 D ★

 C ★

 B ★

 D ★

 A ★

keyAddition # 1

2 + 1 = 3

---

2 + 2 = 4

key #3

 A    B

 B    C

 C   A

Solve the apple puzzle

key Addition #2

3 + 1 = 4

- - - - - - - - - - - - - - - - - - - -

4 + 0 = 4

 A ★

 C ★

 B ★

 B ★

 C ★

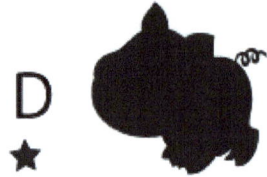 D ★

 D ★

 A ★

HOW MANY....?

Animal	Count
Fox	4
Fish	5
Dolphin	3
Pig	2
Chicken	6
Turtle	1

key Addition #3

3 + 5 = 8

6 + 2 = 8

Key #3

8 7 6

7 8 5

Solve the crocodile puzzle

 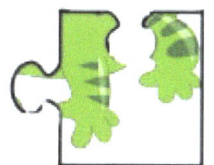

key

Fruit	Number
pear	6
mango	5
strawberry	7
banana	4
cherry	0
blackberry	3

 A ★

 B ★

 B ★

 C ★

 C ★

 D ★

 D ★

 A ★

key #4

3 + 3 = 6

- -

5 + 4 = 9

Solve the unicorn puzzle

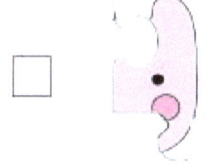

key #4

3 + 4 = 7

4 + 6 = 10

Solve the shape's color

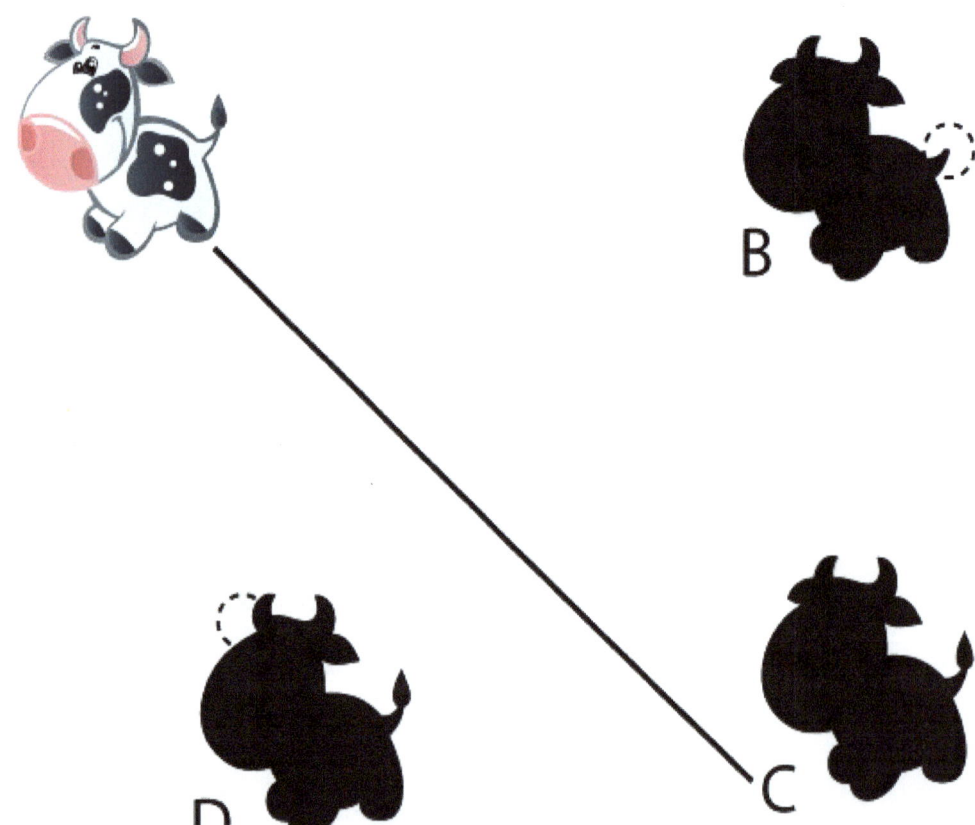

A

B

D

C

solve how many shapes #1

Shape	Count
□	5
▭	7
△	7
☆	6
○	4

key #2

 A  C

 B  D

 C  A

 D  B

solve how many shapes #2

- ☐ 5
- ▭ 7
- △ 7
- ☆ 6
- ○ 4

www.ingramcontent.com/pod-product-compliance
Lightning Source LLC
Chambersburg PA
CBHW051911210526
45473CB00006B/1971